The Algarve Tiger

**Siobhan Mitchell & Eduardo Gonçalves
with António Sabater**

Text copyright © Siobhan Mitchell & Eduardo Gonçalves
Copyright photos © António Sabater

Published by Vista Ibérica Publicações
Urbanização Lagoa-Sol - Lote 1-B - 8400-415 Lagoa
Tel: 282 340660 - Fax: 282 343088 - e-mail: vistaiberica@vistaiberica.com

Design by Peter Daughtrey
Printing by Phalempin, Indústria Gráfica - Porto
I.S.B.N 972-8044-40-2
Depósito Legal Nº 173271/01

The Authors

SIOBHAN MITCHELL has set up a number of conservation and community initiatives in Britain, and is a founder member of SOS LYNX - Associaçao de Defesa da Fauna e Flora - in Portugal, where she now lives. She previously worked for a British national weekly newspaper, and is now one of the best-known English language journalists in the Algarve.

EDUARDO GONÇALVES is Portuguese correspondent for The Sunday Times, The Observer and the Guardian newspapers, and a columnist with The Ecologist magazine. He has written extensively about the wildlife habitats of southern Portugal. Previously he was a political adviser and worked for a number of non-governmental organizations in Britain. He lives in a traditional 'monte' in the south-west Alentejo.

The Photographer

ANTÓNIO SABATER has dedicated his life to capturing images of the magical fauna and flora of the Iberian peninsula. He helped found the prestigious 'Enfoque 10' nature pictures agency and is a winner of the BBC Wildlife Photographer of the Year Award.

The photographs in this book are the result of four years tracking the elusive Iberian Lynx in the wild.

Previous work: 'El Lince Iberico' published by EGMASA, Seville. 1999

Part 1
Getting to know
THE IBERIAN LYNX

Introduction

When we think of rare and exotic big cats, our minds usually wander to the tangles of the Amazonian rainforest, the far-off jungles of Africa, or the stunning mountain peaks of the Asian sub-continent. In fact, the world's most endangered wild cat lives in western Europe. The Iberian Lynx, sometimes called Europe's tiger, inhabits the open forests and aromatic scrubland of south-western Spain and Portugal. If this surprises you, then you will probably be amazed to learn that the largest lynx population in Portugal lives in what most people think of as a holiday destination - the Algarve.

This book will introduce you to the 'tiger' of the Algarve - a creature as cunning as a fox, with the power and grace of a leopard … and a bear's penchant for honey if it gets the chance. The spectacular pictures of the Iberian lynx within this book took photographer António Sabater four years to capture. Prepare yourself for a treat.

But first - just what is an Iberian lynx, and why is it so endangered? One of the world's oldest big cats, it has survived alongside the people of the Iberian peninsula for thousands of years. Today, however, its situation is dire. If nothing is done, there can be little doubt that it will soon be extinct. In 1988, a survey showed there were just 1,200 Iberian lynxes left. Right now, though, there could be as few as 300 of these magnificent animals still roaming the wild. In the last four decades, the areas populated by this unique big cat have shrunk by a staggering 90 per cent. From a species that not long ago covered the entire peninsula (and could even be found in parts of southern France), it is now reduced to inhabiting a handful of tiny, scattered pockets of land.

It is only very recently, after scientists became aware that numbers had declined so sharply, that it has been studied in any detail. It is without doubt an elusive creature - even the most experienced trackers have rarely seen it in the wild - and its ability to camouflage itself in trees or thickets has added to its mystique. This is why it is sometimes referred to as 'the phantom of the forest'. One Algarve resident told recently how she had a strange feeling she was being watched… and moments later caught a fleeting glimpse of a lynx in the thick scrub near her home.

The story of the lynx, though, is also one about the people and traditions of rural Portugal. Its plight is closely entwined with that of local communities. A journey into the rich and dangerous world of the Mediterranean jungle's majestic tiger takes us far into the reaches of what some call the 'real Portugal'…

Siobhan Mitchell & Eduardo Gonçalves
November 2001

Chapter 1

The lynx was revered by ancient societies as a mysterious creature possessing special powers. According to legend, it could see what no man could see. It shares its name with Linceo, the Greek god whose eyes could drill through rock. It was Linceo, according to ancient myth, who guided Jason and the Argonauts to the Golden Fleece.

Descended from an African wild cat that roamed the earth up to 35 million years ago, the Iberian Lynx may be the oldest modern big cat of them all. It is one of over 30 members of the wild cat family, which includes tigers, lions, cheetahs and leopards. Its closest relatives include the American bobcat. It lives in Europe's south-western peninsula, and nowhere else.

The lynx evolved and thrived in the region's oak forests after the Ice Age, developing its palate to the abundance of wild rabbits. The old word for Spain, 'Hispania', is actually thought to mean 'land of rabbits'. The distinctive ears of the lynx may even be an adaptation to its hunting of rabbits. Crouched in long grass or scrub, its dark ear tufts mimic the black tipped ears of the European wild rabbit, its favourite food, so making it easier to get closer to the prey.

The Iberian peninsula is partially detached from the rest of Europe by the Pyrenees mountain range. This helps explain why many species of animal, including the Iberian wolf and the Iberian lynx, evolved separately from other European mammals. The grey wolf and the Eurasian lynx - which is still found in northern and eastern Europe - are both almost twice the size of their Iberian cousins.

Squeezed between the two continents of Europe and Africa, the warm and dry Mediterranean weather coupled with cool humid Atlantic winds created a rich mosaic of different climates, which in turn encouraged a variety of animal and plant species to emerge. The contrast of mountains and plains, forests and steppe, wild rivers and scrubby foothills is home to dozens of mammal species and nearly 400 different birds, many of them exclusive to this region. The Mediterranean region also has an incredible wealth of plants - 25,000 separate species, representing 10 per cent of the world's total number, even though the region only covers 1.6 per cent of the world's surface area.

What makes the area so special is that this rich environment is largely the result of human interaction over many thousands of years, and the farming practices of the area's early settlers favoured the lynx. The use by local people of the forests' natural resources for animal fodder, fuel and wild foods meant that the thick bushy layers under the trees - which the lynx uses for shelter - were preserved and well-managed. The opening of small clearings for grazing and orchards allowed rabbit numbers to grow even further, and created ideal forest edges from where the lynx could launch its attacks on its favourite prey.

Tucked away in the hills are the occasional mud-brick farmhouse, and traditional local breeds of sheep and goats.

Chapter 2

At first glance, the Iberian Lynx looks like an enormous domestic cat. It measures up to three feet long and two feet high. Males weigh between 26-31 pounds, whilst females are usually 20-22 pounds. Its eyes are a piercing green-yellow colour. It lives on average for a little over 10 years - the oldest known animal in the wild died just before its 14th birthday.

The coats of some lynxes may be almost tabby in appearance, with mottled markings on a beige background. Others are virtually bright orange with distinctive black spots, giving it an almost leopard-like look. In fact, *Lynx pardina* - the Iberian Lynx's official scientific name - could be translated as 'Leopard Lynx'. Each coat, though, is as individual as its owner. The lynx also has distinctive tufts on the tips of its ears, a short black, bobbed tail, and a unique 'beard' or ruff that grows with age. These features set it clearly apart from its cousins in the cat world.

The Iberian lynx - in common with other members of the cat family - purrs when contented, growls when threatened, and yowls when it is looking for a mate. Like any domestic cat it also likes to sharpen its claws, but instead of scratching your favourite furniture, the lynx will usually use its favourite tree, often the soft bark of a cork oak tree. It is a skilled climber, and will sometimes choose to rest high in the leafy boughs of a large tree, such as a cork or holm oak, where young lynx are often raised.

It is a solitary animal that comes alive at dusk, choosing to sleep through the heat of the day and use its exceptional powers of night-vision to hunt until soon after the sun rises again. Normally only during the cooler winter months will a lynx stir during daytime. Above all, it plans its timetable to fit with that of its favourite prey, the wild rabbit, which makes up to 95 per cent of its diet.

As a hunter it is highly skilled and well equipped. Its coat gives effective camouflage designed to help it blend into the dappled light of a woodland edge and conceal itself behind bushes as it waits for a rabbit to emerge from its warren. Its hearing is so acute that it can hear a rabbit chewing grass up to 1,000 feet away and detect the slightest rustle from great distances.

It is as a hunter that this animal is undoubtedly at its most impressive. It will watch its prey patiently, tensing its muscles and almost burrowing itself into the ground, waiting for exactly the right moment to pounce. Surprise is its greatest weapon, and it knows that if it is to be successful it must not launch itself too early. Lynx are not long distance runners, and if they do not time their attack to perfection, their prey will have disappeared into a warren or out of reach into the air.

The attack itself is spectacular. From a completely motionless crouch, the lynx will suddenly explode, flying over the bush in one giant leap to quickly asphyxiate its hapless victim with a single bite to the neck. Lynxes have been seen leaping as high as 10 feet into the air in a single bound to catch a red-legged partridge. Mission completed, it will then take its dinner to a secluded spot in the bushes where it can eat undisturbed. If the dish of the day is a partridge, it will first carefully pluck out the feathers before settling down to dine.

A single rabbit a day is enough to satisfy the lynx's hunger, though a female with a young litter may need as many as three rabbits per day. This is one reason why the lynx is an almost exclusively solitary creature - rabbits are too small to share. From time to time (especially when rabbits are in short supply) a lynx may indulge in some of its other favourite foods. As well as red-legged partridges, it also occasionally eats geese, hares, mice and even deer.

Unlike many cats, though, lynxes are not afraid of water, and have been seen jumping into rivers in pursuit of ducks, or swimming to where a dead fish may be floating on the surface. Neither are they afraid to break into a beehive in search of honey and pollen. A lynx, like many human inhabitants of the Iberian peninsula, apparently has a sweet tooth.

Once it has finished its hunting expedition, it will return to its favoured spot among oak trees or bushes within which it seeks refuge from the searing daytime temperatures and hot winds of an Iberian summer. Here, too, it will find the absolute tranquillity it craves.

The lynx is a 'super-predator', which means that it is at the top of the food chain. It also means that it plays a vital role in helping to keep the entire ecosystem in natural balance. It will control the numbers of other, more opportunistic carnivores such as foxes and the Egyptian mongoose, ensuring there are always plenty of rabbits throughout the area.

Chapter 3

A lynx will normally only give up his or her solitary lifestyle during the mating season, which starts at the end of December and lasts until the middle of February. Then, the unmistakable calls of a lusty female on heat may pierce the cold, misty winter nights as she seeks to attract the attention of a nearby male. Females can breed from as young as two years old, although they may not necessarily do so every year. Males are almost always monogamous.

When the female comes into heat, it will be for just three or four days. This will be one of the rare occasions when lynxes may be seen in pairs. A father-to-be will hunt with his new mate, and lovingly wash and groom her.

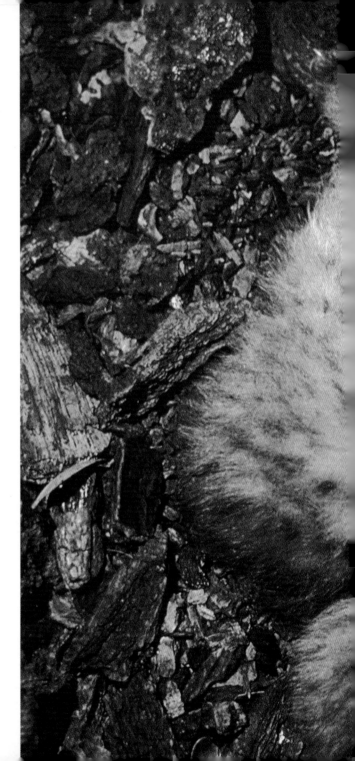

The soft bedding of a hollow cork oak tree makes an ideal den. These three lynx are just a few days old.

Although males play no part in the rearing of the cubs, they may remain in the area after the female has given birth, helping to guard the mother's territory, and that of his new family. Males have also sometimes been observed returning to the area many months later, almost as if hosting a family reunion.

Pregnancy lasts about nine weeks, and during this time the female will scout for suitable dens in which to give birth and raise her litter. These will often be in the hollow of a cork oak tree, although lynx have even been known to raise their brood in old stork nests on the branches of these trees. Where mature oaks are not available, the female will make a nest in dense thicket or among large boulders on a secluded rocky outcrop.

She gives birth between March and May to two or three cubs, although litters of four and even five have been known. Weighing just five ounces, they are born with their eyes closed. Unable to control their body temperature, the cubs rely on their mothers to keep them warm. She stays with them 24 hours a day for the first few days of their lives, drawing on energy reserves from her final hunting trips before the birth. After about four days, she has to leave them in search of food so that she can continue to provide milk. Her hunt must be brief, however, as she cannot afford to leave the vulnerable young cubs for long.

The litter grows rapidly, weighing about 1 lb. after two weeks, and nearly 2 lb. at one month. It is about now that the mother moves them to a second den in order to avoid detection by potential predators, and ensure that they are not infested by undesirable parasites. The old den also becomes increasingly small for the boisterous, blue-eyed cubs. Their mother picks up the cubs by the scruff of the neck and hides them one by one in nearby bushes. They may stay there for only a few days before the mother moves them on again. In the first few months of their lives, young lynxes may come to know as many as ten different homes.

Now out in the open for the first time, the cubs, with a cat's typical curiosity, begin to explore their intriguing new surroundings. They also play with their brothers and sisters.

It is during these games of hiding, crouching and pouncing that these young wild cats hone the skills they will need in order to survive and hunt for themselves. Their mother joins in these games and also brings back small, live mammals.

This is a difficult time for the mother, who needs to catch two or three rabbits a day in order to feed both herself and her young family. Often, one of the cubs does not survive this period. The mother carefully buries the body under branches and earth in an almost human ritual.

The surviving cubs soon begin to accompany their mother on hunting trips and must now learn to curb their youthful enthusiasm and keep quiet. Their first attempts to hunt result in, at best, a few feathers, or the sight of a rabbit fleeing into its warren. Cubs are also likely to come across the other strange creatures that share their world, discovering that hedgehogs have spikes that hurt your nose and that toads do not taste so good after all. With time and an experienced hunter to teach them, though, they are soon able to feed themselves.

When they are about six or seven months old, the fast-growing lynx cubs start to become more independent of their mother, although they remain within her territory and at first only venture further out in the company of a brother or sister. However, the time is fast approaching for a cub to leave and find a territory of its own. A female cub may be lucky and inherit that of her mother, or even share it with her. For a male, though, finding a home of his own will be the most important - and dangerous - experience of his life.

The young male begins to make brief excursions outside his mother's territory from about eight or nine months, although it may be only at one or even two years of age that he makes the final break. He travels dozens of miles and faces many obstacles and threats. Sadly, studies have shown that the majority - as many as 80 per cent - die during the journey, usually on a busy road or in a hunter's trap.

If he survives these obstacles, he must quickly find an area that has both sufficient

food and shelter. He must also have water, not just to quench his thirst on hot summer days, but because rivers and their tributaries are used by lynxes to navigate their travels much as we use roads. Once established, they use these landmarks to travel around their territory and to favoured hunting areas.

The ideal territory a young lynx is seeking ranges in size from a few hundred to as many as 3,000 acres, depending on available resources, the most important of which is the number of rabbits. A male may deliberately overlap his home range with that of a female to make sure he has a partner with which to reproduce.

If he is lucky, he will be able to take over a territory left by a lynx that has died recently. Otherwise, if he stumbles onto one that is already occupied and there are no other suitable areas nearby, he may try to force the resident lynx to flee. He will start a war of attrition that may last several months, first gathering intelligence on the current occupier by sniffing its droppings. These include what are called 'lynx stones' - the hardened urine pellets unique to this creature - which a territory-holder sprays on bushes along the edges of his land to show who owns it. The younger lynx will also inspect tree scratches, to determine how old and strong his potential adversary is.

As competition for scarce habitats increases, a number of lynxes have been found killed or seriously injured after territorial fights. And because of human disturbance - hunting, logging or new development - it is now becoming harder for lynxes to permanently settle a territory. These animals may spend much of their lives as wandering vagrants. It is then that they are most vulnerable to snares, or even to unscrupulous hunters' bullets.

Diseases and the destruction of habitats result in some young lynxes being forced to return home hungry to their mother's territory, to build up strength before trying again. Others, however, are unable to find food during the long journey and simply starve.

For a lynx that makes it through the dangerous early days and successfully establishes its own territory, life is now mainly spent in its own company. The daily search for food is its most important priority. Its everyday routine follows a familiar pattern, usually starting with a trip to its favoured watering hole for a refreshing drink. It also periodically tours and scent-marks its territory in order to keep potential challengers away.

For the greater part of the day, however, it behaves like many a domestic cat, sharpening its claws and resting in a carefully chosen spot where it can watch but not be seen. It will prove its reputation as 'the phantom of the forest,' rarely emerging from its well-camouflaged dens within its territory - until, that is, the mating season comes around and the sharp yowls of the female lynx once again pierce the Iberian night air.

Part 2
Living with THE LYNX

Chapter 4

It is perhaps hard to imagine, but the lynx once lived in or very near Lisbon and Madrid. Fossils and bones have been discovered close to both of today's Iberian capital cities, giving us an idea of how different these areas must have been before they became urban centres. The people of Spain and Portugal have co-existed with the Iberian lynx for literally thousands of years. For most of this time, human impact on this wild creature has been minimal, yet in very recent years humans have been responsible for bringing this animal to the very brink of extinction.

The mixed farming system of the "Montado" is recognised as the world's most biodiverse.

25

In ancient times, the Iberian peninsula was relatively teeming with lynx. We know this because the people of the Neolithic era made charms out of their teeth and claws, and occasionally even ate their meat. Around 20,000 years ago, paintings of these creatures were etched in caves by our prehistoric ancestors.

As early humans evolved in Europe, we changed from nomadic hunter-gatherers to settled farmers. The Mediterranean region is said to be 'the cradle of civilisation' because the earliest evidence of agriculture can be found here, dating back more than ten thousand years. These small steps towards a new lifestyle were to have the greatest impact on the Iberian peninsula's wilderness area, shaping what are today's forests, scrubland and meadows, and helping to create here what is now one of the most biodiverse regions in the world.

Initially the attempts to grow food, rather than just gather wild plants and hunt animals, would have meant clearing small patches of forest in order to allow light to reach crops. The first animals to be domesticated - the ancestors of today's grazing animals such as goats and cows - would also have needed more open land to graze. These grasslands are today home to many types of plants, animals and birds that are dependent on the traditional methods of cultivation which maintain them.

These first steps towards agriculture benefited the lynx by increasing the numbers of rabbits, as these thrived on the grassy pastures and fields of grain. By controlling rabbit numbers, the lynx was an early ally of farmers, who would otherwise suffer large-scale crop losses. Moreover the Iberian lynx, unlike larger carnivores such as the wolf, would not have posed a threat to farm animals.

Although human populations expanded and opened up new areas for farming, these changes occurred slowly over a long period, left large expanses barely touched and, by and large, allowed wildlife to adapt alongside us. In recent years, though, the rate and ferocity of change to the landscape has shattered this equilibrium.

The first major blow to affect the lynx, however, was an accident.

*The meadows within the woods are ideal for
rabbits - and for hidden lynx to hunt in.*

Chapter 5

In the 1950s, a French doctor introduced a disease called myxomatosis into his garden to rid himself of rabbits that had damaged his vegetable patch. It was a move that was to have unforseen and catastrophic repercussions. The fatal disease soon spread far beyond his own back yard, wreaking havoc on wild rabbit populations throughout Europe. In 'Hispania', in what was once the land of the rabbit, they were virtually wiped out. In neighbouring Portugal, where myxomatosis arrived in 1956, rabbit mortality rates reached 95 per cent. For the lynx, which had adapted as a specialist feeder, it was disaster.

At around the same time, rural landscapes were beginning to change in both countries. In Portugal, the wheat campaign - which sought to turn the countryside into the nation's breadbasket - meant large areas of forest and scrub were destroyed. By the 1950s, the thickets which had covered as much as 90 percent of the Algarve and Alentejo hills were reduced to as little as 10 per cent in some areas, such as the Serra do Caldeirão.

The campaign itself was a failure because of poor quality soils and erosion caused by attempts to grow cereals intensively. The poverty that followed caused people to leave rural communities in large numbers and abandon land which then deteriorated to dense scrub, benefiting only the wild boar.

These rapid changes to the landscape, coupled with the disastrous collapse of rabbit numbers, were not only devastating for the lynx. They were serious, too, for birds of prey such as the imperial eagle. They helped cause the extinction, as recently as the late 1960s, of the black vulture and the wolf in the Algarve.

In the last 30 years, new and more intensive farming methods have been introduced into many of these same areas. Thousands of acres of land that was once covered with scrub or forest have been cleared for greenhouse-grown fruit, vegetables and flowers, or for row upon row of eucalyptus and pine plantations. Neither of these forms of farming leave much room for wildlife. Nor do the large dams that have been built to irrigate these intensive crops and the many golf courses for which the Algarve is justifiably renowned.

As Spain and Portugal developed into more prosperous societies, tarmac surfaced roads replaced dirt tracks which could only be travelled over on foot or on a donkey. Highways were built through previously untouched areas, creating virtual barriers between neighbouring lynx populations. Motor traffic brought new and deadly dangers.

Nowadays, mass tourism is posing an increasing threat to the lynx as new holiday villages spring up, sometimes illegally, in what were previously mature woodlands. In Portugal, the area south of Lisbon has come under particular pressure with the building of the new bridge across the River Tagus and a motorway linking the capital with the Algarve. Thousands of mature oak trees have been felled to make way for new developments along this route and more fellings are expected.

The increasing popularity of off-road sports is bringing motor rallies into the heart of lynx country in the Algarve and Alentejo regions, and even into the lynx reserve of Malcata in the north-east of Portugal. For a lynx seeking peace and tranquillity, hundreds of 4x4 vehicles and trial motorbikes roaring through your forest is hardly entertainment.

Lynx were recently seen here - but this area will soon be submerged by Europe's biggest reservoir.

Chapter 6

The mountains of Monchique, which straddle the northern Algarve and southern Alentejo regions, are the last hope for the lynx in Portugal. But here, too, the process of change is exacting a heavy toll.

Most developments we see here today did not exist even a few decades ago. For instance, the main road from Portimão on the coast up through the mountains to the town of Monchique was built only in 1936, after which the latter became a trading centre for the region. Very few people are thought to have lived in these mountains until one thousand years ago and the arrival of the Moors. It was they who developed the first irrigation networks and built the first tracks through the mountains making it possible to grow food all year round. Many terraces with stone walls up to five metres in height were created during this era and fed by natural springs through channels cut into the granite rock They are still maintained and in use today.

Around five hundred years ago Portugal developed an empire and new foods were brought in from overseas. Fruits such as loquats from China, and vegetables such as potatoes from South America were introduced to gardens for the first time. Much of the mountain range, however, would still have been inaccessible and rarely used by people, except for the seasonal gathering of berries and firewood, or by goat herders.

It was only around fifty years ago that major change came to the region when the then Salazar dictatorship decided to 'develop' rural Portugal. By building new tarmac roads, lives were revolutionised almost overnight. What had once been a three-day journey by donkey now became a 30-minute journey by car. Suddenly the rest of the outside world was within easy reach. These early roads to the tiny mountain villages were used to transport crops to market as well as bring in fertilisers to aid their growth. Products such as cement were also brought in so that, for the first time for many people, the earthen floors of their cottages were replaced with a solid surface. Other products became available, such as plastic pipes to carry water directly to people's homes and newly-built terraced gardens.

Serra da Brejeira: a lynx stronghold on the Algarve-Alentejo border.

Above all, though, this new access to the outside world drew many young people away from these villages, lured by the promise of a more comfortable life in cities, holiday resorts and overseas. It is a drift from the land which continues to this day throughout rural Portugal and Spain. In many villages, only the elderly now remain.

Cars, tractors and lorries could now reach the mountainsides. Soon the first paper and pulp companies moved in to buy or lease land from farmers in order to plant eucalyptus on the steep slopes. Eucalyptus grows particularly well in this climate and has one of the shortest planting to cutting cycles in the world. It is also very profitable, especially where land is cheap and plentiful through abandonment. Nowadays, more than half of the Monchique mountain range is covered with eucalyptus trees, many of them planted during the 1970s and the 1980s at the expense of native woodlands and scrub.

These plantations leave a sweet smell on the breeze, but for wildlife it is a very different story. The dry and barren slopes provide little food or shelter for lynxes, genets, wild cats or otters. The tranquillity of the region is frequently shattered by the noise of tractors and chainsaws as trees are felled and trucked out of the hills to paper factories further north.

The thirst of these alien trees for water has dried many local springs and streams, making even small-scale farming virtually impossible. The steep slopes, cleared of their natural vegetation, are prone to erosion and mudslides during the heavy winter rains. During the hot summer months, the risk of uncontrollable fires running through the closely-planted and fast-burning trees hangs constantly in the air.

The descendants of the first Iberian lynx which settled in the tranquil scrub and forests of Monchique's mountains have today inherited just small pockets where they can barely survive and must cope with new hazards posed by fast traffic and the ever-present threat of forest fires. Once abundant sources of water and wild rabbits have virtually disappeared. The few lynxes that remain are rarely seen and, even then, often only after they have been accidentally shot or trapped. There was a recent report here of a lynx attacking a lamb, the scarcity of their natural prey bringing them for the first time into direct conflict with their ancient natural ally, the small farmer.

HUNTERS AND THE LYNX

Whilst our ancient ancestors revered the lynx as a skilled hunter, it would rarely have been hunted itself. The development of the fur trade during the last century, however, led to thousands of lynxes being killed. In Spain during the 1930s, hundreds of lynx pelts were traded each year. Later, after the Second World War, the lynx was included on the list of 'vermin' to be eradicated from the countryside. The hunting of the animal became popular among the privileged classes.

It was not until as recently as the mid-1970s that the lynx became a protected species in both Spain and Portugal. Nevertheless, many are still shot - some at close range where there could have been little doubt in the hunter's mind as to the identity of his target. Many of those killed are young and inexperienced lynx away from their mothers for the first time in search of food or a territory to call their own.

Although poaching for fur is now virtually non-existent, the lynx is mistakenly seen as a threat to scarce rabbit populations, the hunting of which is a profitable sport for many country estates. In fact, studies have shown clearly that in the absence of lynxes other

carnivores, which take a much higher number of rabbits, quickly spread. In turn, the resultant scarcity of rabbits increases the chances of a lynx being accidentally caught in traps as hunters are forced to set more of them.

Organised culls of other wild animals such as the wild boar also pose a risk to the lynx. In theory, fox and wild boar culling could benefit the lynx as both animals compete with it for scarce rabbits. In practice, however, these hunts - with as many as 300 dogs taking part - commonly occur during the lynx's breeding season or when females are due to give birth. Pairs of lynx have been seen fleeing from such hunts and it is hardly surprising that those unable to escape have been killed on several occasions.

Hunting nowadays has become a more widespread leisure-time activity. It is no longer the preserve of the aristocracy, or a means of capturing food. New large estates where people pay to hunt wild animals are increasing in popularity in both Portugal and Spain. These could, some argue, help in the preservation of wild landscapes, which in turn could benefit the lynx and its prey. In the absence of rabbits, though, many now focus on large game such as deer. As a result, habitat management practices have been changed, slowing the recovery of rabbit populations in the wild. High fences built to keep the deer in also pose an insurmountable barrier to lynx in search of food or a mate.

Part 3
The missing
LYNX

Chapter 7

In a few short years, modern society has reversed centuries of living in relative harmony with the lynx. In the process, it has virtually destroyed an invaluable natural heritage of the kind thought to exist only in distant rainforests.

As a result, the Iberian Lynx is a vanishing species. Its numbers are falling dramatically. In the 1960s, it was thought there were approximately 3,000 lynxes. By the end of the 1980s, this had fallen to 1,200. In 1996, new research suggested the figure had halved. Now, there are fears there may be just 300 left, huddled into isolated pockets totalling just 10 per cent of the size of their 1960s territories. The phantom of the forest, it seems, may soon be just a ghost in our memories.

There are around fifty groups still surviving, but more than half the remaining population live in just two areas: in the Spanish hills of Sierra Morena and Montes de Toledo. Of the remainder, many have only a handful of animals and are not likely to survive for long. Even the larger groups with relatively stable populations face a mounting number of threats to their very existence. The latest evidence suggests that even in Sierra Morena and Montes de Toledo, numbers may be falling again. As many as a quarter of the remaining lynx groups face immediate threats to their habitats.

In spite of their protected status, traps, snares and bullets are still the leading cause of non-natural mortality, claiming up to 85 per cent of animals - some in vicious, outlawed snares. In each case, the victims are usually young animals and their early death will have deprived populations of desperately needed new genetic resources. Even if a lynx survives, it may be left badly injured and unable to hunt. In recent years, many lynxes have been found with mutilated legs and broken teeth, the result of their struggle to free themselves. Poisons laid for other mammals also pose a threat. Like traps, poisons rarely discriminate between their victims.

Large areas of natural vegetation, which have taken thousands of years to evolve, have

been obliterated. In their place have come industrial-scale plantations for logging and intensive agriculture. More woodlands have been deliberately burnt to make way for urban development, or bulldozed to make way for new highways and dams. Vast plantations of pine and eucalyptus now occupy large areas of south-western Spain and Portugal where old native woodlands and bush once stood. The forced terracing of hillsides with heavy machinery and the dense planting of these alien trees have caused an environmental disaster. Soil has crumbled away and water sources have dried up. What is left is virtual desert, a sterile, lifeless area with no plants or bushes, let alone wildlife.

The eucalyptus was virtually non-existent in the Iberian Peninsula up until the 1960s. Now it occupies almost as much land as the indigenous cork oak tree, for which Portugal and Spain - the world's largest producers of cork - are traditionally famous. Many plantations have been installed with generous European Union subsidies, which in turn come from the pockets of European taxpayers, but the cost to wildlife species like the lynx is immeasurable. The frequent logging of these trees means tracks are opened up into the heart of wilderness areas, forcing lynx to flee.

The plantations have also increased both the number and ferocity of forest fires in a region where summers are notoriously long, hot and dry. In the last 10 years, over two million acres of woodland have been burnt in Portugal and there are 8,000 forest fires every year in Spain. Those lynx which survive are usually driven to hostile environments where there may be little food - and new dangers. Yet the number and area of plantations are continuing to grow, as are the EU funds to install them.

The Guadiana River divides Portugal and Spain - this is a 'crossing point' for the Iberian Lynx.

The problem has been aggravated by the decline of traditional farming practices, seen by officials and planners as backward and unprofitable. The resultant rural exodus has created a vicious circle: as land is abandoned, local resources lose value and are commonly replaced by plantations. However, these degrade the soil and drain groundwater, which means local resources lose even greater value, prompting even more people to leave the land.

For the lynx, the abandonment of farms can mean more than the wholesale substitution of its territories. The rabbit's habitat also deteriorates as neglected pastures become overgrown. There is an increasing trend among wealthy city-workers to buy up old farms and convert them to weekend homes. As farms and entire villages close, increasingly popular off-road motor sports move in.

There are currently plans to build a number of new dams and highways in key lynx areas. Already, several dams spanning tens of thousands of acres have destroyed entire ecosystems once inhabited by the lynx. Many river valleys rich in rabbits, and along which lynxes moved in search of a mate, have been inundated by water. In at least one case, the Alqueva dam in eastern Portugal, the clearance of scrub and woodland for Europe's largest reservoir has taken place during the breeding and rearing season in a place where lynxes have been repeatedly seen in recent years. As with plantations, such developments are made possible by generous European subsidies, even when they contravene the EU's own conservation laws. And like plantations, they pose a formidable barrier to a lynx as it travels in search of a new territory. Motorways and other highways are bringing heavier and faster traffic into lynx areas. Many have already become 'black spots' for the animal. They, too, look set to increase in number.

Just as rabbit populations began to develop immunity to myxomatosis, disaster struck again. In the 1980s and early 1990s, a new disease - haemorreghic viral disease - once again virtually eradicated rabbits from the Iberian Peninsula. Today, rabbit numbers are just five per cent of former levels. Without rabbits, the lynx as a species cannot survive. A female will only breed when there are enough rabbits to support her and her cubs. A male will only settle in an area if rabbit resources are plentiful. There can be few more shocking sights than that of a starving, emaciated lynx barely able to walk, let alone hunt. Yet such tragic spectacles have recently been captured on film.

As the dwindling populations of the Iberian Lynx are scattered into small and more isolated pockets, these groups become vulnerable to any setback - the failure of a single female to breed, for instance, or the death of a young lynx on a road. The entire species becomes more prone to disease as lynxes are forced to breed with close relatives, because there are no other potential mates nearby. Already, scientists have found evidence of massive gene abnormalities in the sperm of male lynxes. Some fur patterns have disappeared, and there have been sightings of 'melanistic' or black lynx, evidence of the appearance of a recessive gene, thought to be due to in-breeding. As today's lynx populations become smaller and weaker, the possibility of a domino effect provoking a sudden collapse of the entire species has become a serious prospect.

A pile of freshly cut cork in this lynx habitat corridor.

Part 4.
What hope for THE IBERIAN LYNX?

Chapter 8

The Iberian Lynx is now in what scientists term a 'pre-extinction phase'. This means that without urgent action, it will most likely die out in just a few short years. Many of the animals that remain are little more than refugees, travelling ever greater distances in desperate search of food and shelter, encountering more and more dangers and unable to find a mate. Sightings of pairs and young cubs are now rare. The disappearance of each population group, or the elimination of one more connecting 'corridor' between lynx groups, becomes another giant leap towards eventual extinction. However, the end for the Iberian Lynx is not inevitable. It is only if nothing is done that it will slide into the history books.

Conservation programmes have been drawn up by agencies of the Spanish and Portuguese governments, but so far have yet to be fully implemented. Many important lynx habitats have been listed for inclusion in the EU's 'Natura 2000' network of protected wildlife areas, but it may still be many years before this scheme comes into force. Research and monitoring programmes are under way in both countries and tentative efforts have been made to increase rabbit populations in some areas. But much more must - and can - be done.

If the lynx is to have a long-term future, the traditional sustainable methods of mixed farming in those areas it still inhabits must find a new voice. Known as *dehesas* in Spain and *montados* in Portugal, they have evolved from their ancient roots into a mosaic of open forest, pasture and fruit and vegetable gardens from which many people still meet their needs locally, as well as providing a rich and varied habitat for many types of birds and other wildlife.

The extensive cork oak forests provide a vital income to local communities from the cork bark that is stripped every nine years using a method which has remained almost

unchanged for the last 3,000 years. The wine industry's continuing demand for natural wine bottle corks has thus preserved entire communities and ecosystems. Holm oaks are pruned to provide charcoal, and the sweet acorns that fall from these trees provide food for the famous black pigs of Portugal and Spain, from which delicious smoked hams are produced.

The strawberry tree shrub produces bright red berries, which are harvested in rural areas to make *medronho,* an alcohol used in local taverns to warm smoked sausage or drink neat on cold winter mornings. Bees find ample pollen on the flowers of herbs, such as lavender, which grow in undisturbed forests and scrubland. You can still see in use traditional beehives made of a complete cylinder of cork bark. Cattle, sheep and goats are still grazed extensively amongst the forests and meadows, and are the source of a wide variety of extraordinary fresh and smoked cheeses.

The fruits of this idyllic reality could and should be revalued. In an increasingly consumer-conscious world, many of us now seek foods produced naturally. A whole array of labels promoting such values adorn the shelves of stores, confirming the ready demand for such products. There are a number of other forestry products which could be harvested sustainably and which would add enormous value to areas of scrub which planners see as 'unproductive'. By rewarding farmers for continuing to use sustainable practices, the lynx could still enjoy a long-term future. And by investing in these communities, local people benefit - as does the environment and wildlife.

The hills inhabited by the 'Algarve Tiger' are a treasure trove of wild, perennial foods and medicinal plants, some of whose properties we have probably yet to discover. The berries of the strawberry tree are perhaps the most concentrated source of vitamin C available. Amazingly, this potential has yet to be developed. Tests have also shown that valuable essential oils could be extracted from rock rose bushes which grow profusely all over the hillsides, but local communities have been unable to exploit this potential through lack of investment. The traditionally sustainable use of natural landscapes produces a whole range of high quality foods - such as honey, olive oil, cheeses and wild mushrooms - that bring an important income for local people and are the basis of the famous, healthy Mediterranean diet. But the lack of access to outside markets means such delicacies go untasted and their potential unexplored.

As each day goes by, these areas face new threats. Without a true recognition of their value, even these few remaining lynx habitats could now go. Demand from the paper industry for fast-growing, quick-profit wood pulp is increasing. Logging companies have privately earmarked more than one million acres for new plantations to match projected market growth. Obligingly, more generous subsidies are being made available to aid this expansion. At the same time, the advent of plastic wine stoppers has cast a dark cloud over the thousands of people who

depend directly on the cork forests - not to mention the wildlife species such as the lynx for whom these forests are their last refuge.

If the lynx is to make it through just the next five years, urgent action is needed to tackle problems such as rabbit scarcity, for example by creating breeding centres to produce ample supplies of disease-free wild rabbits for release in lynx areas. The reintroduction of lynx bred in captivity into areas where numbers are so low they may not have a wide enough gene pool to sustain themselves is now surely inevitable. So far, however, neither measure has got underway.

In the spring of 2001 a group of individuals - including conservationists, biologists and wildlife experts - got together to launch a new charitable organisation called *Associação de Defesa de Fauna e Flora* and a campaign called SOS LYNX. The Iberian Lynx has reached its current situation because it is virtually unknown, and hence forgotten and neglected. It is the intention of SOS LYNX to ensure that what is probably the biggest current conservation issue in Europe is put at the top of the agenda. This book represents the start of that process. There will be little pressure on governments and politicians to conserve a species no-one has ever heard of. There can be no excuse now.

It is also the intention of SOS LYNX to work directly on the ground and in local communities to help people preserve - and promote - the unique ecosystems they have created and on which wildlife like the lynx depend. This means we will be urging consumers to buy, and companies to invest in, products and sustainable farming systems that enhance the quality of lynx habitat. And it is also the intention of SOS LYNX to help restock rabbit populations and stabilise lynx numbers, including through land purchases and lobbying, for which we will need your help.

We believe that the fate of the Iberian Lynx is not merely a conservation issue, or one about animal rights or welfare. It touches on wider questions about the path society is taking and its consequences for all of us. The threat of the first-ever extinction for over 2,000 years of a big cat living on our doorstep should be a big wake-up call. It should tell us that, somewhere along the line, we have gone badly wrong in our quest for economic growth and what is commonly called 'progress'. This is a debate we cannot avoid if we wish to avoid future extinction crises.

But for now, we do not have long to set things right, at least in so far as the Iberian Lynx is concerned. An emergency situation demands emergency measures. It is not too late to stop the 'Tiger of the Algarve' from disappearing forever.

But the time is now. Please join us.